# GARBUZ
# SPACE SCHOOL
# ACADEMIES

## FOLLOW IN THE FUTURE WITH THE SHADOW

## GEORGIY SERGEYEVICH GARBUZ

Published 2025

Printed in the United States of America

First Edition
ISBN (softcover): 978-1-963380-66-8
ISBN (hardcover): 978-1-963380-67-5
ISBN (e-book): 978-1-963380-68-2

For information, address:
Holzer Books LLC
8 The Green, Ste. A
Dover, Delaware 19901 USA

For information about special discounts available for bulk purchases, sales promotions, and educational needs, contact:
info@holzerbooksllc.com
+1 (888) 901-7776

**holzerbooks**LLC®

# Contents

Part IV

Impact on the Space Industrial Revolution

Part V

Social and Humanitarian Impact

# INTRODUCTION

# Chapter 1

# Setting the Stage

T he future of humanity depends on its ability to uplift every child, especially those who have been left behind. Among the most overlooked and underserved groups in societies worldwide are children in foster care—individuals brimming with untapped potential but too often denied access to opportunities that could transform their lives.

## The Current State of Education for Foster Care Children

Around the world, millions of children live in foster care systems, placed there due to family crises, abuse, neglect, or abandonment. These children face a range of educational challenges that often prevent them from reaching their full potential. While education systems aim to serve all students equally, foster care children experience significant disadvantages:

1. **Frequent School Transfers**: Moving between homes often results in disrupted schooling, missed instruction, and a lack of educational continuity.

2. **Resource Inequality**: Many foster care children lack access to proper materials, mentorship, or programs tailored to their needs.

3. **Psychological and Emotional Barriers**: Trauma, instability, and societal stigma weigh heavily on their ability to focus on education and imagine a future beyond their current circumstances.

4. **Limited Support Systems**: Without strong family structures, foster care children often lack role models or guidance to help them pursue higher education or careers.

The result is a tragic reality where foster children graduate at significantly lower rates, enter adulthood unprepared for stable careers, and are disproportionately affected by poverty and societal neglect. Left unsupported, they are denied the chance to contribute meaningfully to their own lives and to society as a whole.

## The Challenges They Face

The challenges faced by foster care children extend far beyond the classroom. These barriers are deeply rooted in systemic inequalities and social neglect, limiting their opportunities to pursue careers in high-demand, competitive fields like science, technology, and engineering.

1. **Lack of Opportunities**: Without access to specialized programs, hands-on learning, and exposure to cutting-edge technology, children in foster care are excluded from fields such as aerospace, engineering, and space sciences.

2. **Societal Neglect and Marginalization**: Society often fails to recognize or prioritize the needs of foster care children, resulting in minimal policy focus or funding for tailored educational initiatives.

3. **Limited Life Choices**: Without meaningful educational and career guidance, many foster children face bleak outcomes, including unemployment, homelessness, or dependence on welfare systems.

These challenges do not reflect a lack of talent or ability within foster care children. Instead, they highlight a failure of systems to nurture their potential and provide pathways for success.

## The Revolutionary Vision of Georgiy Sergeyevich Garbuz

It is against this backdrop of systemic neglect that the visionary **Georgiy Sergeyevich Garbuz** emerges with a transformative solution: the creation of the **National Garbuz Space Academy** and a global network of **Garbuz Space School Academies**. These institutions are not just schools—they are lifelines for children who have been ignored and underestimated for far too long.

1. **The National Garbuz Space Academy**:
   • Located in the United States, this will be the largest and most advanced space school academy in the world.
   • A central hub for space education, research, and development, it sets the gold standard for preparing students for careers in the space industry.

2. **The Global Garbuz Space School Academies**:
   • Established in every city or state of every country that is part of the **International Intergalactic Space Federation (IISF)**, these academies ensure that no child, regardless of their location or background, is left behind.
   • By focusing on science, technology, engineering, mathematics, and health sciences, the academies equip students to become astronauts, engineers, scientists, and leaders of the future.

Georgiy Sergeyevich Garbuz's vision is revolutionary because it redefines education as the foundation of a better world—not just for individuals, but for humanity as a whole. By prioritizing foster care children and other underserved groups, Garbuz creates a model of equity and opportunity that bridges the gap between disadvantage and achievement.

This initiative is not merely about education. It is about empowering the youth who will lead us into the next great frontier: **space**. It is about building a society where talent and determination are nurtured, regardless of a child's origin.

The **National Garbuz Space Academy** and **Garbuz Space School Academies** represent a bold new future—one where foster care children and underserved youth are given the tools they need to reach for the stars. Through this visionary program, education becomes the great equalizer, transforming lives and ensuring that every child has the opportunity to become a vital part of humanity's mission to explore, colonize, and innovate in the cosmos.

This is more than a dream; it is a promise: *A promise to uplift the forgotten, to inspire the next generation of space pioneers, and to build a united, thriving future for all.*

# Chapter 2

# The Core Vision

T he **Garbuz Space School Academy Initiative** is more than an ambitious educational program—it is a transformative vision for humanity's future. Driven by a commitment to equity, innovation, and global collaboration, this initiative aims to bridge the gap between Earth's underserved youth and the infinite possibilities of space exploration. At its heart are two pillars: the **National Garbuz Space Academy**, the flagship institution in the United States, and a global network of **Garbuz Space School Academies** in every city or state across participating nations of the **International Intergalactic Space Federation (IISF)**.

## The National Garbuz Space Academy

The **National Garbuz Space School Academy** stands as the centerpiece of this visionary project. Designed to be the largest and most advanced space school academy in the United States and the world, this institution will serve as the epicenter of space education, technological research, and workforce development.

1. **A Symbol of Excellence**:
   • The National Garbuz Space Academy will set a new global standard for space-focused education, housing state-of-the-art laboratories, simulation centers, and research facilities.
   • Students will gain hands-on experience with advanced technologies, including

robotics, artificial intelligence, and space medicine.

2. **Capacity and Impact**:
   • Accommodating tens of thousands of students annually, the academy will nurture future astronauts, space scientists, engineers, and medical professionals.
   • Its graduates will serve as pioneers, leading humanity's efforts to explore, colonize, and sustain life in space.

3. **Innovation Hub**:
   • The academy will act as a hub for cutting-edge research and development, collaborating with the world's top space agencies, governments, and private industries.
   • Technologies developed here will drive advancements in both space exploration and life on Earth, addressing challenges such as energy sustainability, healthcare, and climate change.

4. **Inspiration for the World**
   • As the flagship institution, the National Garbuz Space Academy will inspire similar academies worldwide, becoming a beacon of progress, hope, and opportunity for all.

## The Global Garbuz Space School Academies

While the National Academy serves as the central hub, the **Garbuz Space School Academies** represent the initiative's far-reaching impact on a global scale. These academies will be established in **every city or state of every country** that is part of the **International Intergalactic Space Federation (IISF)**, ensuring that education is accessible to all, regardless of geography or socioeconomic background.

1. **Global Reach, Local Impact**:
   • By establishing academies in every city or state, the program guarantees that no child is left behind. Whether in large metropolises or small rural towns, students will have access to high-quality, space-focused education.
   • Each academy will adapt to local needs while maintaining a standardized

curriculum to ensure global alignment.

2. **Fostering Global Unity**:
• The academies will unite diverse cultures and nations under a common goal: preparing humanity for its next great journey into space.
• Students from all backgrounds will collaborate, building bridges across borders and fostering a sense of shared purpose and understanding.

3. **Career Preparation for Space and Beyond**:
• The academies will focus on STEM disciplines—physics, biology, mathematics, engineering, and health sciences—to equip students for careers in space industries.
• Graduates will be prepared for roles as astronauts, space doctors, scientists, engineers, researchers, and other critical positions in the global space economy.

4. **Massive Capacity for Change**:
• The program will serve **30,000 students per country annually**, amounting to **5,910,000 students globally**.
• Each year, **1,182,000 graduates** will enter the workforce as highly skilled professionals ready to contribute to space exploration and global development.

## Aligning with Broader Goals: Space Exploration, Education, and Global Development

The Garbuz Space School Academies align seamlessly with humanity's larger goals of advancing space exploration, improving educational systems, and fostering sustainable global development.

1. **Empowering the Workforce for Space Exploration**:
• The academies address the growing demand for skilled professionals in the aerospace, medical, and engineering industries.
• By nurturing talent on a global scale, the initiative ensures that humanity has the workforce needed to sustain lunar bases, establish Martian colonies, and explore the deeper cosmos.

2. **Promoting Education as the Foundation for Progress**:
   • Education is at the core of this vision. By providing advanced training and cutting-edge resources, the academies empower youth to break free from cycles of poverty and societal neglect.
   • The initiative elevates education to a tool for global equality, ensuring that every child—no matter their background—has the opportunity to thrive.

3. **Global Collaboration for a Better Future**:
   • As part of the **International Intergalactic Space Federation (IISF)**, the Garbuz Space School Academies symbolize what humanity can achieve through cooperation.
   • By uniting nations in a shared purpose, the program promotes peace, cultural exchange, and a vision of a more unified Earth.

4. **Innovations That Benefit Earth**:
   • Research and technologies developed through the academies will address critical challenges on Earth, from energy sustainability to medical breakthroughs.
   • Space exploration serves as a catalyst for innovation that improves life on our home planet while opening pathways to the stars.

The **National Garbuz Space Academy** and the global network of **Garbuz Space School Academies** represent a transformative vision for humanity's future. By creating opportunities for millions of young minds worldwide, this initiative prepares the next generation of space leaders, scientists, and pioneers. It aligns education with humanity's collective goals of exploration, collaboration, and progress, ensuring that Earth's children will be the architects of our journey into the cosmos.

# Part 1

## The Origins of the Garbuz Vision

# Chapter 3

# The Inspiration

At the heart of the **Garbuz Space School Academies** lies a deeply personal story, shaped not only by the life of **Georgiy Sergeyevich Garbuz** but also by the enduring influence of his parents, **Sergey Georgiyevich Garbuz** and **Lyubov Vasilievna Garbuz**. Their values, resilience, and unwavering dedication to creating opportunities for others left an indelible mark on Georgiy, inspiring him to pursue his visionary mission of transforming education and empowering children worldwide.

## Georgiy's Childhood in Kazakhstan

Born and raised in Kazakhstan, Georgiy grew up in a region marked by both natural beauty and economic hardship. In the shadow of the Baikonur Cosmodrome—one of the world's most significant space launch sites—Georgiy developed a fascination with space at an early age. He watched as rockets soared into the sky, symbolizing humanity's ambition to explore beyond Earth. Yet, on the ground, he also witnessed a stark contrast: children left behind, struggling against poverty, societal neglect, and a lack of opportunity.

In Kazakhstan's foster care system, Georgiy saw children with immense potential—bright, curious, and eager to learn—fall through the cracks of an underfunded and underprepared educational system. Many were shuffled between homes and schools, unable to build the stability they needed to thrive. Their dreams were stifled not because of their abilities, but because society had failed to nurture their talents.

This experience left an indelible mark on young Georgiy. He recognized the injustice of a world that could launch humans into space while ignoring children who had no chance to rise above their circumstances. It planted in him a deep conviction: *education could be the bridge that lifted these children from poverty and helped them achieve extraordinary things.*

## Breaking Generational Cycles of Poverty Through Education

For Georgiy Sergeyevich Garbuz, education is more than a tool for learning—it is the key to transforming lives, families, and entire societies. He firmly believes that **education is the only way to break generational cycles of poverty**, providing children with the skills, knowledge, and opportunities to escape the limitations of their past.

1. **A Pathway to Opportunity**:
   Education gives children a chance to dream bigger and aspire to careers that seemed impossible. It opens doors to professions like science, engineering, and medicine—fields that empower individuals to contribute meaningfully to society while improving their own lives.

2. **A Tool for Equality**:
   Georgiy understands that talent and potential are universal, but opportunities are not. By focusing on underserved children—particularly those in foster care—he seeks to level the playing field, ensuring that no child is denied the chance to reach for the stars.

3. **A Foundation for Humanity's Future**:
   Georgiy's vision extends beyond individual success. He believes that educating children today lays the foundation for a better world tomorrow. By investing in the next generation, we prepare humanity to solve its greatest challenges, explore new frontiers, and build a brighter future.

## A Personal Mission

Georgiy's childhood experiences taught him that neglecting a child's education is not just a personal tragedy—it is a loss for all of humanity. Each child represents untapped potential, a future leader, a scientist, a creator, or an innovator who could change the world if only given the chance.

This understanding has driven Georgiy to dedicate his life to creating opportunities for foster care children and underserved youth. The **Garbuz Space School Academies** are a direct result of his belief that education can unlock the potential of every child, no matter their background. These academies are not just schools; they are lifelines, designed to lift children out of poverty and prepare them to lead humanity's journey into the cosmos.

Through his vision, Georgiy Sergeyevich Garbuz honors the children he grew up alongside—children whose potential was never realized—and ensures that future generations will have the opportunities they deserve.

The story of the Garbuz Space School Academies begins with Georgiy's own story—a story of witnessing injustice, recognizing potential, and dedicating himself to creating change. His belief that **education is the key to breaking poverty** and unlocking human potential serves as the driving force behind this revolutionary initiative.

# CHAPTER 4

# THE 2007 TURNING POINT

I n 2007, a convergence of economic, political, and scientific factors set the stage for one of the most transformative ideas in modern history: the creation of the **National Garbuz Space Academy** and the global network of **Garbuz Space School Academies**. For **Georgiy Sergeyevich Garbuz**, this year marked a turning point—not only for his personal vision but for the future of education, technology, and humanity's exploration of space.

## The Economic and Political Climate: A Time for Bold Ideas

The mid-2000s brought significant global shifts that paved the way for revolutionary concepts like the Garbuz Space School Academies. By 2007, the world stood at a unique crossroads:

1. **Economic Recovery and Growth**:
   - After overcoming earlier recessions, economies around the globe began experiencing renewed growth. Governments and investors were increasingly looking for bold, high-impact projects that could fuel technological progress and create opportunities for the next generation.
   - Space exploration, once seen as an unattainable luxury, regained attention as

advancements in technology made it more accessible and cost-effective.

2. **Political Interest in Global Innovation**:
   • Nations began recognizing the need to invest in education and science to maintain economic and geopolitical influence in a rapidly changing world. The idea of training a new workforce for space-related careers became not only practical but necessary for global competitiveness.
   • The growing spirit of international collaboration, particularly through organizations like the early foundations of the **International Intergalactic Space Federation (IISF)**, created opportunities to align nations under shared goals.

3. **Societal Focus on Education**:
   • Education was increasingly seen as the cornerstone of economic development and social progress. Governments, businesses, and philanthropists began championing programs that focused on STEM (Science, Technology, Engineering, and Mathematics) to address future workforce demands.

In this environment of growth and innovation, visionary leaders like Georgiy Sergeyevich Garbuz found both the inspiration and the opportunity to turn ambitious dreams into actionable plans.

## Georgiy's Revolutionary Science Projects and Medical Breakthroughs

While the world prepared for its next era of advancement, **Georgiy Sergeyevich Garbuz** was already several steps ahead. By 2007, Georgiy had devised groundbreaking contributions that could change industries and save countless lives. These achievements became the financial and intellectual foundation for the **Garbuz Space School Academy Initiative**.

1. **Medical Breakthroughs**:
   • Georgiy pioneered revolutionary treatments in regenerative medicine, including therapies that repaired damaged tissue at the cellular level and advanced cures for previously untreatable diseases.

- These breakthroughs could bring global recognition and secure significant revenue streams, as governments, medical institutions, and private organizations invested in his research to improve healthcare systems worldwide.

2. **Space-Related Innovations**:
- Georgiy's work extended beyond Earth's boundaries into technologies that had direct applications for space exploration. From sustainable life-support systems to anti-radiation shields for astronauts, his innovations solved critical challenges in interstellar travel.
- These projects could attract international partnerships and funding from aerospace agencies, further cementing Georgiy's reputation as a pioneer of both science and space development.

3. **Financial Resources and Vision**:
- Revenue that could be possible generated from Georgiy's scientific discoveries, patents, and medical treatments can become the primary funding source for his ambitious educational vision.
- Instead of focusing solely on profit, Georgiy plans to direct his wealth toward initiatives that aligned with his belief in education as the key to societal progress.
- He saw the opportunity to create lasting change: a program that would leverage the advancements of science to lift up foster care children, underserved youth, and humanity as a whole.

## The Birth of a Vision

The alignment of the global climate and Georgiy's revolutionary breakthroughs brought his ideas into sharp focus. In 2007, the groundwork was laid for the creation of:

1. **The National Garbuz Space Academy**:
- The flagship academy in the United States, envisioned as the largest and most advanced space school in the world. This institution would serve as a hub of research, innovation, and education, producing graduates capable of shaping humanity's future in space.

2. **The Global Garbuz Space School Academies**:
   • With a clear mission in place, Georgiy developed a blueprint to establish space-focused academies in every city of every country within the **International Intergalactic Space Federation (IISF)**.
   • This network would provide millions of children, particularly those from foster care and underserved communities, with the education and skills needed to become space pioneers.

3. **A Unified Vision for Humanity**:
   • Georgiy's projects were not just about education; they were about creating a global community united by science, exploration, and opportunity. His vision aimed to prepare Earth's youth for the next great leap: the colonization of space and the development of sustainable extraterrestrial civilizations.

The year 2007 became a turning point not only for Georgiy Sergeyevich Garbuz but for the world. It was the year bold new ideas took root, nurtured by an economic and political climate that embraced innovation and driven by Georgiy's revolutionary breakthroughs in science and medicine.

This convergence of factors allowed Georgiy to transform his vision into reality: the creation of the **National Garbuz Space Academy** and a global network of **Garbuz Space School Academies**. By investing his achievements into education, Georgiy laid the foundation for a future where every child—no matter their background—could reach for the stars and lead humanity into a new era of exploration and discovery.

# Chapter 5

# The Founding Principles

At the heart of the **National Garbuz Space Academy** and the global network of **Garbuz Space School Academies** lie the guiding principles set forth by **Georgiy Sergeyevich Garbuz**—a powerful and simple philosophy captured in the words:

**"Za Detey – For Kids! Za Lyubov – For Love! Live, Make, & Enjoy!"**

These words are more than a slogan; they are the foundation upon which this revolutionary initiative is built. They reflect Georgiy's unwavering belief that education, love, and purpose are the keys to a brighter future for humanity.

## "Za Detey – For Kids!"

The first and most fundamental principle is a dedication to the children—the leaders, dreamers, and explorers of tomorrow. Georgiy Sergeyevich Garbuz believes that the future of humanity starts with empowering the youngest generations, particularly those who have been overlooked and underserved.

1. **Prioritizing Foster Care and Underserved Youth**:
   - The Garbuz Academies place special emphasis on children from foster care and marginalized communities, offering them opportunities that have historically

been denied.

• Georgiy recognizes that every child, regardless of their circumstances, has the potential to contribute to society and reach extraordinary heights.

2. **Empowering Through Education**:

• By focusing on science, technology, engineering, mathematics (STEM), and health sciences, the academies equip students with the skills and knowledge to pursue careers in the growing space economy.

• Education becomes a tool for transformation, enabling children to rise above poverty, societal neglect, and other barriers to success.

3. **Creating Equal Opportunities**:

• The principle "For Kids!" reflects a commitment to leveling the playing field so that no child is left behind.

• The global network ensures that access to quality education is not limited by geography, nationality, or socioeconomic status.

## "Za Lyubov – For Love!"

Love is the driving force behind Georgiy Sergeyevich Garbuz's vision: love for children, love for humanity, and love for the possibilities that lie beyond Earth. This principle speaks to the compassion, hope, and unity at the core of the Garbuz Space Academies.

1. **A Vision Rooted in Compassion**:

• Georgiy's personal experiences witnessing the struggles of foster children inspired a deep sense of responsibility to create a better world for them.

• "For Love!" reminds us that this initiative is built on care and empathy, driven by the desire to provide children with opportunities for fulfillment and happiness.

2. **A Global Mission of Unity**:

• Love transcends borders, and so does the mission of the Garbuz Academies. By establishing schools in every city of every IISF-member country, this initiative fosters cooperation and cultural exchange among nations.

• Students learn not just to succeed as individuals but to collaborate, creating a global community united by shared purpose and mutual respect.

3. **Love for Exploration and Humanity**:

• Georgiy's vision extends beyond Earth. "For Love!" reflects the belief that exploration is an act of hope and a demonstration of humanity's limitless potential.

• It is through love for discovery and the unknown that humanity can unite to explore the stars and secure a future for generations to come.

## "Live, Make, & Enjoy!"

The final principle emphasizes action, creativity, and joy. It is a call to live with purpose, to create boldly, and to celebrate the achievements that arise from hard work and collaboration.

1. **Live with Purpose**:

• The Garbuz Academies inspire students to live meaningful lives filled with curiosity and ambition.

• By preparing students for careers in space and STEM fields, the academies provide a clear path toward purposeful contributions to society.

2. **Make a Better Future**:

• "Make" represents the act of creation—whether it's building new technologies, solving global challenges, or establishing sustainable space colonies.

• The Garbuz Academies empower students to become innovators, leaders, and builders of humanity's next great chapter.

3. **Enjoy the Journey**:

• Georgiy's philosophy reminds us that life is to be enjoyed. Education should inspire excitement, discovery, and a sense of wonder.

• The academies foster a supportive and engaging environment where students can pursue their passions and celebrate their successes.

## How These Principles Drive the Garbuz Academies

The founding principles—**"For Kids! For Love! Live, Make, & Enjoy!"**—are not mere words; they are the guiding force behind every aspect of the National Garbuz Space School Academy and its global network:

1. **Child-Centered Education**:
   • Programs are tailored to the needs of students, particularly foster care children and underserved youth, ensuring that education is accessible, inclusive, and empowering.

2. **Global Impact**:
   • Schools in every IISF-member country reflect the principle of "For Love," promoting unity and cooperation on a worldwide scale.
   • The global reach ensures that millions of children are given the tools to live purposeful lives and shape humanity's future.

3. **Purpose and Innovation**:
   • The focus on space sciences and cutting-edge technologies aligns with "Make," encouraging students to contribute to groundbreaking innovations that benefit humanity both on Earth and in space.

4. **A Celebration of Humanity**:
   • The academies inspire students to dream big, live fully, and enjoy the process of learning, discovery, and creation.

The founding principles—**"Za Detey – For Kids! Za Lyubov – For Love! Live, Make, & Enjoy!"**—are the soul of Georgiy Sergeyevich Garbuz's vision. These words reflect his deep belief in education as the key to unlocking human potential, his compassion for children who have been left behind, and his unwavering hope for a united and prosperous future.

Through the **National Garbuz Space Academy** and the **global network of Garbuz Space School Academies**, these principles come to life, creating a world where every child has the opportunity to live, learn, create, and explore—not just for themselves, but for the benefit of all humanity.

This vision is not just about building schools—it is about building futures, fostering love, and inspiring generations to reach for the stars.

# Part II

## The Garbuz Space School Academy Network

# CHAPTER 6

# THE NATIONAL GARBUZ SPACE ACADEMY

The **National Garbuz Space Academy** is the centerpiece of Georgiy Sergeyevich Garbuz's revolutionary vision for education and humanity's future in space. As the **largest space-focused educational institution in the world**, it stands as a flagship for innovation, research, and the preparation of a new generation of space pioneers. Located in the United States, the academy serves as both an educational powerhouse and a symbol of global collaboration and progress.

## The Largest in the World: A Flagship for Advanced Space Education

The **National Garbuz Space Academy** is unparalleled in its scope and scale. Designed to be the most advanced institution of its kind, it serves as a **global beacon** for excellence in STEM education (Science, Technology, Engineering, and Mathematics), with a specialized focus on preparing students for careers in the space industry.

1. **Global Leadership in Space Education**:
   - The academy sets a **gold standard** for space education, integrating cutting-edge technologies, hands-on research, and real-world applications.
   - By hosting the brightest minds, educators, and scientists, it will lead the way in training future astronauts, space doctors, engineers, scientists, and other critical

roles in space exploration.

2. **Comprehensive Facilities**:
   • The academy features **state-of-the-art laboratories**, space simulation centers, research hubs, robotics workshops, and observatories.
   • Students will have access to facilities such as zero-gravity simulators, advanced planetary habitat replicas, and AI-driven engineering labs to prepare for space environments.

3. **Capacity to Change the World**:
   • With the ability to educate **tens of thousands of students annually**, the National Academy is unmatched in scale, providing opportunities to youth from all backgrounds.
   • Its graduates will not only fulfill the workforce demands of the rapidly growing space economy but also inspire innovation across Earth's industries.

# A Central Hub for Research, Teacher Training, and Innovation

Located in the United States, the National Garbuz Space School Academy is positioned as the **central hub** for global space education, technological innovation, and teacher development.

1. **Space Research and Development**:
   • The academy will host interdisciplinary research programs, tackling the greatest challenges of space exploration, such as sustainable life-support systems, advanced propulsion technologies, and interstellar communication networks.
   • Partnerships with space agencies, governments, and private-sector leaders will drive innovations that advance humanity's ability to explore and colonize other worlds.

2. **Training Future Educators**:
   • The National Academy will serve as a **training ground for teachers** who will staff the Garbuz Space School Academies around the globe.
   • Educators will undergo rigorous training programs, learning to implement

cutting-edge teaching methods, integrate AI-driven learning tools, and inspire students to excel in STEM fields.

3. **A Center for Collaboration**:
• The academy will bring together scientists, engineers, researchers, and policymakers from around the world to develop solutions for both Earth-based challenges and interstellar exploration.
• International conferences, workshops, and collaborative missions will be hosted at the academy, solidifying its role as the heart of global space education and innovation.

## Capacity: Tens of Thousands of Students Prepared for the Future

The scale of the **National Garbuz Space School Academy** reflects the magnitude of its vision. The academy will educate **tens of thousands of students annually**, offering a rigorous, future-focused STEM curriculum tailored to the demands of the modern space economy.

1. **Rigorous STEM Curriculum**:
• The curriculum combines foundational STEM education with specialized training in areas such as: space medicine, robotics and AI applications for interstellar missions, planetary resource extraction and sustainability, advanced physics and astrophysics, aerospace engineering and zero-gravity systems, students will graduate with not only theoretical knowledge but also hands-on experience, ready to enter the workforce as contributors to humanity's exploration of the cosmos.

2. **Opportunities for All**:
• The academy prioritizes accessibility for students of all backgrounds, particularly foster care children and underserved youth, aligning with Georgiy Sergeyevich Garbuz's mission to uplift those left behind.
• Scholarships, mentorship programs, and partnerships with international governments will ensure that talented students, regardless of socioeconomic status, can access this world-class education.

# A Symbol of Humanity's Commitment to a Space-Ready Generation

The **National Garbuz Space School Academy** represents more than an educational institution—it is a **symbol of hope, progress, and humanity's collective ambition**.

1. **Building a Global Space-Ready Workforce**:
   - By educating the brightest minds from around the world, the National Academy prepares a generation ready to address the challenges of space exploration and interplanetary settlement.
   - Its graduates will form the backbone of the workforce that will drive the International Intergalactic Space Federation's (IISF) missions to the Moon, Mars, and beyond.

2. **Inspiring the World**:
   - As the largest and most advanced space academy in existence, it will inspire nations, educators, and youth to see education as a pathway to progress and possibility.
   - The academy's success will encourage governments to invest in similar initiatives, creating ripple effects of innovation and development worldwide.

3. **A Monument to Human Potential**:
   - The National Academy stands as a testament to what humanity can achieve when vision, collaboration, and compassion are combined.
   - It symbolizes Georgiy Sergeyevich Garbuz's belief that *every child has the potential to reach the stars*, and it represents humanity's commitment to ensuring that opportunity is available to all.

The **National Garbuz Space Academy** is more than the largest space school in the world; it is a cornerstone of humanity's future. As the flagship institution for advanced space education, innovation, and research, it serves as a model for excellence and a beacon of hope for millions of children.

With its unparalleled capacity, groundbreaking curriculum, and global impact, the National Academy represents a bold step toward a future where **every child has the opportunity to dream, create, and explore—on Earth and beyond.**

# Chapter 7

# Garbuz Space School Academies Worldwide

Building on the success of the **National Garbuz Space Academy**, the **Garbuz Space School Academies** represent the next phase of Georgiy Sergeyevich Garbuz's transformative vision for global education and space exploration. These academies will ensure that no child, regardless of their location or background, is denied access to a world-class education that prepares them for humanity's future among the stars.

## Global Reach: A Presence in Every Country

The Garbuz Space School Academies will be established in **every city or state of every country** that is part of the **International Intergalactic Space Federation (IISF)**. This global presence ensures that the initiative touches the lives of millions, empowering youth worldwide with the tools and skills necessary to thrive in the space economy.

1. **Universal Accessibility**:
    - By establishing academies in both urban centers and smaller communities, the program guarantees that all children—regardless of geographic location—can benefit from advanced STEM education.
    - This global reach fosters inclusivity, breaking down barriers to opportunity and leveling the playing field for underserved and marginalized populations.

2. **Cultural and Economic Impact**:

• Each academy will serve as a hub of knowledge and innovation within its local community, driving advancements in science, education, and workforce development.

• The academies will also act as a symbol of international cooperation, fostering a sense of unity among nations working together under the IISF.

## Annual Capacity: Educating the World

The scale of the Garbuz Space School Academies reflects their bold mission: to prepare millions of students for careers that will drive the space economy and support interplanetary exploration.

1. **Per-Country Impact**:

• Each participating country will host Garbuz Space School Academies capable of educating up to **30,000 students annually**.

• This ensures significant national impact, creating a pipeline of highly skilled graduates ready to contribute to space industries and technological advancements.

2. **Global Scale**:

• Collectively, the Garbuz Space School Academies will educate an astonishing **5,910,000 students globally** each year.

• This massive educational effort represents the largest coordinated initiative for STEM education in human history.

3. **Annual Graduates**:

• Of the total student body, **1,182,000 students will graduate annually**, equipped with the skills, knowledge, and experience to enter the global space workforce.

• These graduates will form the backbone of the workforce required to sustain humanity's ambitions in space, including lunar settlements, Martian colonization, and deep-space exploration.

## Standardized Curriculum: Preparing Students for the Cosmos

The curriculum of the Garbuz Space School Academies is designed to provide a rigorous and future-focused education, aligned with the demands of the space economy.

1. **Core STEM Disciplines**:
   Students will receive a strong foundation in:
   • **Physics**: Focused on space systems, gravitational dynamics, and propulsion technologies.
   • **Biology**: Addressing life sciences in space environments, human adaptation to microgravity, and sustainable agriculture for extraterrestrial colonies.
   • **Mathematics**: Emphasizing advanced calculations, data analysis, and applied problem-solving for space missions.
   • **Engineering**: Covering aerospace, robotics, and the design of life-support systems and planetary habitats.
   • **Health Sciences**: Preparing future space doctors and researchers to address the unique medical challenges of living and working in space.

2. **Hands-On Learning**:
   • Students will engage in practical, project-based learning that mirrors real-world challenges in space exploration.
   • Simulations, robotics competitions, AI development, and resource extraction labs will allow students to apply their knowledge in dynamic and innovative ways.

3. **Global Standardization with Local Relevance**:
   • While the curriculum will maintain a consistent global standard, it will be adapted to local cultural and educational systems.
   • Partnerships with regional educators and space agencies ensure that each academy meets the unique needs of its community while preparing students to thrive in a global context.

## Career Preparation: Building the Space Workforce of Tomorrow

The Garbuz Space School Academies are designed to prepare students for critical roles in the emerging global space economy. Each graduate will possess the skills and experience to succeed in a wide range of space-related professions, including:

1. **Astronauts**:
   • Students will train in life sciences, physical fitness, and mission operations, with advanced preparation for roles as astronauts leading human missions to the Moon, Mars, and beyond.

2. **Space Doctors**:
   • Specialized programs in space medicine will equip students to address the unique health challenges of space exploration, such as radiation exposure, zero-gravity health risks, and mental well-being during long missions.

3. **Engineers and Technologists**:
   • Graduates will excel in fields like aerospace engineering, robotics, AI, and planetary infrastructure, contributing to the development of space habitats, propulsion systems, and life-sustaining technologies.

4. **Space Scientists and Researchers**:
   • Focused on areas such as astrophysics, planetary science, and resource extraction, these professionals will advance humanity's understanding of the universe and develop tools for sustainable extraterrestrial living.

5. **Support and Innovation Roles**:
   • From mission planners to space-based agricultural experts, students will graduate ready to innovate and contribute across diverse roles critical to sustaining humanity's expansion into space.

The **Garbuz Space School Academies** are more than schools; they are **launchpads for humanity's future**. With their global reach, standardized curriculum, and focus on STEM excellence, these academies will prepare millions of young minds to lead humanity's journey into the stars. By empowering students with knowledge, skills, and purpose, Georgiy Sergeyevich Garbuz's vision ensures that the opportunities of the space economy are accessible to all—no matter their background or geography.

Together, these academies will form the foundation of a **space-ready generation**, uniting the world in the pursuit of knowledge, exploration, and progress.

# Chapter 8

# The Role of Local Governments

The success of the **Garbuz Space School Academies** hinges on a collaborative effort between the **International Intergalactic Space Federation (IISF)**, the global network of schools, and the governments of participating nations. Local and national governments play a critical role in ensuring the sustainability, accessibility, and growth of the academies, providing the foundation upon which these institutions can thrive. By leveraging infrastructure, resources, and partnerships, governments will help realize the vision of preparing millions of students for the space economy.

## Collaborating with National Governments to Provide Infrastructure, Technology, and Funding

1. **Infrastructure Development**:
   Local governments will work closely with the IISF to identify, build, or repurpose facilities that meet the high standards required for Garbuz Space School Academies. This includes ensuring:
   • Access to **modern classrooms and laboratories** equipped with cutting-edge technology.
   • Development of **simulation centers** for practical training in zero-gravity

environments, robotics, and resource extraction.

• Construction of facilities such as **observatories** and research hubs where students can engage in real-world science.

• Governments will oversee land allocation and zoning processes to facilitate the construction of academies in urban, suburban, and rural regions, ensuring geographic accessibility for all communities.

2. **Technological Support**:

Governments will help integrate emerging technologies into the schools to align with the demands of space-focused education. This includes:

• Providing **high-speed internet access** and digital learning platforms to ensure students and teachers can connect globally and collaborate effectively.

• Implementing **AI-based tools** for personalized learning, research simulations, and curriculum development.

• Ensuring access to **advanced software** for engineering, robotics, and planetary science studies.

3. **Financial Contributions and Funding Models**:

Governments will collaborate with the IISF and private sector partners to secure funding for the establishment and operation of the academies. Sources of funding include:

• **National education budgets**: Allocating funds specifically for space education initiatives.

• **Public-Private Partnerships (PPPs)**: Collaborating with corporations, universities, and international organizations to finance infrastructure and program development.

• **Tax-based contributions**: Governments may direct revenue from industries benefiting from space research and education to support the academies.

• Financial models will prioritize sustainability, ensuring that Garbuz Space School Academies remain operational for future generations.

4. **Monitoring and Governance**:

• Local governments will play a role in monitoring the progress of the academies, ensuring quality education standards and equitable access to programs.

• Government-appointed advisory boards may oversee the integration of local needs into the standardized curriculum and ensure transparency in operations.

## Utilizing Public Facilities for Education Programs and Outreach Initiatives

To make education widely accessible and foster community engagement, local governments will support the use of existing public infrastructure to complement the operations of Garbuz Space School Academies.

1. **Leveraging Public Facilities**:
   Governments will enable the use of **community centers, libraries, and educational institutions** for supplementary programs such as:
   • Evening classes and weekend workshops to accommodate students from underserved or rural areas.
   • Public outreach initiatives to inspire interest in space education among families and communities.
   • Temporary facilities may be established in existing public buildings during the construction phase of permanent Garbuz academies.

2. **Outreach and Engagement Programs**:
   Local governments will partner with the academies to host:
   • **Space Education Festivals**: Community events showcasing student projects, space simulations, and lectures by space scientists.
   • **Workshops and Competitions**: Programs that expose students to robotics, coding, engineering, and planetary sciences, encouraging participation from broader communities.
   • **Guest Lectures**: Collaboration with local schools to invite professionals, astronauts, and researchers to inspire students and highlight the possibilities of careers in the space economy.

3. **Scholarship and Enrollment Support**:
   • To ensure the academies are accessible to all, governments will work with the IISF to provide **scholarships** for talented but financially disadvantaged

students.

• Outreach programs in schools, foster care systems, and rural communities will identify and encourage gifted students to pursue education at Garbuz academies.

4. **Integration with Local Educational Systems**:
   • Garbuz academies will complement local education systems by offering advanced courses and training not otherwise available.
   • Governments will facilitate partnerships between Garbuz academies and local high schools or universities, allowing students to access specialized learning paths while remaining within their communities.

## The Collaborative Vision

The role of local governments is essential in turning the vision of the Garbuz Space School Academies into a global reality. Through partnerships in infrastructure, funding, and outreach, governments will ensure that these academies thrive in every city or state of every IISF-participating nation.

This collaboration reflects the core values of the Garbuz initiative: a commitment to providing opportunities for **all children**, fostering global unity, and preparing humanity to explore the cosmos. By working hand in hand with the IISF, local governments can empower students to rise to their full potential, creating a generation ready to lead humanity's next great chapter among the stars.

# Part III

## Implementation and Requirements

# Chapter 9

# Strategic Partnerships

The success of the **Garbuz Space School Academies** relies on a web of strategic partnerships that bring together the expertise, resources, and vision of diverse stakeholders. From the **International Intergalactic Space Federation (IISF)** to private investors, space agencies, and educational institutions, these collaborations ensure that the academies fulfill their mission to prepare a global workforce for humanity's interstellar future.

## Collaboration with the IISF, International Governments, Private Investors, and Educational Institutions

1. **The Role of the IISF:**

   As the primary governing body for interstellar exploration and development, the IISF serves as a central partner in the Garbuz initiative. The IISF provides overarching support, including:

   • **Funding Allocation**: Directing resources from space-related taxes, international grants, and revenue generated by space industries.

   • **Global Coordination**: Ensuring that the academies align with the broader goals of IISF-member countries and fostering cooperation among participating nations.

• **Standardization**: Helping establish a standardized curriculum and benchmarks for space-focused education across all academies.

2. **Partnerships with International Governments**:

Governments of IISF-member countries play a vital role in funding, infrastructure development, and policy support for the academies. These collaborations include:

• **Policy Integration**: Aligning national education policies with the objectives of the Garbuz academies to streamline implementation.

• **Resource Sharing**: Providing land, facilities, and technological resources to support academy operations.

• **Scholarship Programs**: Co-funding scholarships for underprivileged and foster care children to ensure equitable access to education.

3. **Engagement with Private Investors**:

The initiative actively seeks partnerships with private investors and corporations to supplement government and IISF funding. Key contributions from private entities include:

• **Financial Support**: Direct investments in infrastructure, technology, and operational costs.

• **Innovation Funding**: Supporting research initiatives at the academies, particularly in cutting-edge fields like AI, robotics, and sustainable space systems.

• **Public-Private Partnerships (PPPs)**: Collaborating with businesses to create internship programs, mentorship opportunities, and career pipelines for graduates.

4. **Collaboration with Educational Institutions**:

Universities and research institutions are critical partners in shaping the academic rigor and research output of the Garbuz academies. Partnerships focus on:

• **Curriculum Development**: Co-creating advanced courses and programs tailored to the demands of the space economy.

• **Teacher Training**: Providing resources and expertise to train educators who will lead Garbuz academy classrooms.

• **Joint Research Initiatives**: Partnering on projects that address challenges in space exploration, such as resource extraction, life support, and human adaptation to extraterrestrial environments.

## Partnering with Space Agencies and Industries to Align Curriculum Goals with Workforce Needs

1. **Collaboration with Space Agencies**:

Partnerships with leading space agencies, such as NASA, ESA, Roscosmos, ISRO, CNSA, and others, ensure that the academies' curriculum is directly aligned with real-world demands. Key areas of collaboration include:

• **Training Programs**: Developing hands-on learning modules based on the requirements of ongoing and future space missions.

• **Access to Expertise**: Hosting guest lectures, workshops, and mentorship programs led by astronauts, engineers, and researchers.

• **Simulation and Practical Learning**: Providing students with access to cutting-edge simulation technologies used in mission planning and astronaut training.

2. **Engagement with Space Industries**:

Collaboration with private aerospace companies such as SpaceX, Blue Origin, Boeing, and others bridges the gap between education and industry. These partnerships focus on:

• **Internships and Career Pipelines**: Creating internship opportunities and hiring pathways for Garbuz academy graduates.

• **Workforce Training**: Aligning the curriculum with the specific skill sets required for roles in aerospace manufacturing, propulsion system design, robotics, and mission operations.

• **Joint Projects**: Partnering on research initiatives that benefit both the academies and the space industry, such as resource utilization, sustainability, and interplanetary logistics.

3. **Industry-Driven Curriculum Alignment**:

The academies work closely with both public and private space agencies to identify workforce trends and evolving needs. The curriculum is designed to prepare students for high-demand roles, such as:
• Aerospace engineers and propulsion system specialists.
• Space doctors and biomedical researchers.
• AI and robotics developers for planetary exploration.
• Planetary resource extraction experts.
• Space mission planners and coordinators.

4. **Long-Term Workforce Development**:
   • By aligning curriculum goals with industry needs, the academies ensure a steady pipeline of qualified professionals who can meet the challenges of space exploration and colonization.
   •This collaboration positions IISF-member countries as global leaders in the space economy, driving innovation and economic growth.

Strategic partnerships are the lifeblood of the **Garbuz Space School Academies** initiative. Through collaboration with the IISF, international governments, private investors, educational institutions, and space agencies, these academies will create a seamless bridge between education and industry.

By aligning the curriculum with workforce needs and fostering international cooperation, the Garbuz academies will prepare a generation of skilled professionals who are ready to lead humanity into its next great frontier: the cosmos. This unified effort ensures that the dream of space exploration becomes a reality, not just for a few, but for all.

# CHAPTER 10

# REQUIREMENTS FOR SUCCESS

T he establishment and sustained operation of the **National Garbuz Space Academy** and the global network of **Garbuz Space School Academies** demand a robust foundation of resources, talent, and infrastructure. These requirements are critical to ensuring that the initiative achieves its vision of empowering millions of students worldwide and preparing them for careers in the space economy.

## Financing: Securing Funds for the Vision

The financial foundation of the Garbuz Space School Academies relies on a diverse range of funding sources. Securing these funds ensures the construction, operation, and continuous improvement of the academies on a global scale.

1. **Directed Taxes**:
   • Revenue from taxes on scientific discoveries, space-related technologies, and industries benefiting from the initiative (e.g., aerospace, AI, and renewable energy) will provide a stable financial base.
   • Contributions from IISF-member governments ensure long-term sustainability.

2. **International Grants**:
   • Grants from global organizations, such as the United Nations and IISF, as well as international education funds, will support the academies' establishment in developing countries.
   • These grants prioritize equitable access to education for underserved populations.

3. **Private Sector Investments**:
   • Partnerships with corporations in aerospace, technology, and finance will yield significant investments in infrastructure and program development.
   • Investors gain access to a highly skilled workforce and innovative research outcomes, creating a mutually beneficial relationship.

4. **Philanthropic Contributions**:
   • Global philanthropists and foundations dedicated to education and STEM development will be encouraged to support scholarships, special programs, and facility expansion.

## Professional Workforce: Recruiting Top Talent

Building an effective workforce of educators, scientists, and specialists is essential to delivering the high-quality education and training envisioned for the academies.

1. **Qualified Teachers**:
   • Recruitment focuses on STEM educators with advanced degrees and teaching experience in physics, biology, mathematics, engineering, and health sciences.
   • Teachers will undergo additional training at the **National Garbuz Space Academy** to learn cutting-edge teaching methodologies and the integration of space-related topics into the curriculum.

2. **Research Scientists**:
   • Experts in space sciences, robotics, AI, and planetary engineering will lead research programs and mentor students in advanced topics.
   • These scientists will also contribute to collaborative research projects with IISF

and private partners.

3. **Space Specialists**:
• Professionals with experience in astronaut training, mission planning, and space systems will provide hands-on instruction in specialized fields.
• These specialists ensure that students are trained in real-world scenarios, aligning their skills with workforce demands.

4. **Global Recruitment**:
• The academies will attract talent from around the world, promoting diversity and fostering a collaborative, inclusive environment.
• Partnerships with leading universities and space agencies will help identify and recruit top educators and specialists.

## Technology and Materials: Equipping the Academies for Success

The Garbuz Space School Academies require access to cutting-edge tools and materials to provide students with a world-class education.

1. **Classroom Technology**:
Classrooms will be equipped with advanced interactive learning tools, such as:
• AI-driven teaching assistants for personalized instruction.
• Virtual and augmented reality (VR/AR) systems for immersive learning experiences.
• High-speed internet and collaborative platforms for global connectivity.

2. **Laboratory Tools**:
Laboratories will feature state-of-the-art equipment for experiments in physics, biology, chemistry, and engineering. Specialized labs will include:
• Zero-gravity simulators.
• Robotics workshops.
• AI development environments.
• Planetary resource extraction simulators.

3. **Educational Materials**:

• Curricula will be supported by comprehensive digital resources, textbooks, and simulation programs aligned with IISF standards.

• Partnerships with leading publishers and technology providers will ensure access to the latest educational content.

4. **Sustainability and Innovation**:

• Green technologies, such as solar-powered facilities and energy-efficient systems, will reduce the environmental footprint of each academy.

• The use of locally sourced materials ensures cost-efficiency and community engagement in construction and operation.

## Infrastructure: Building the Academies of the Future

Creating a global network of Garbuz Space School Academies and expanding the facilities at the National Garbuz Space Academy requires substantial infrastructure investment.

1. **Construction of Global Academies**:

• Each IISF-member country will host multiple academies, strategically placed to ensure accessibility for all regions.

• Facilities will include classrooms, laboratories, observatories, simulation centers, and recreational areas.

2. **Expansion of the National Academy**:

• The **National Garbuz Space Academy** will serve as a central hub, requiring continual upgrades to accommodate advanced research, educator training programs, and student capacity.

• Additional wings for research and innovation will house partnerships with international space agencies and industries.

3. **Community Integration**:

• Academies will incorporate public spaces, such as libraries, conference centers, and community outreach facilities, to foster local engagement and collaboration.

4. **Scalability and Future Expansion**:

• Infrastructure will be designed with scalability in mind, ensuring that the academies can grow as demand increases.

• Modular construction methods and adaptable designs will allow for rapid deployment in underserved areas.

The success of the **Garbuz Space School Academies** depends on meeting these critical requirements: securing sustainable financing, recruiting top talent, providing cutting-edge technology, and building state-of-the-art infrastructure. By addressing these needs, the initiative ensures its ability to transform global education and prepare a generation of students to lead humanity into its interstellar future.

This vision is a call to action for governments, private partners, and global citizens to unite in building the foundation for a better, more inclusive, and space-ready world.

# Chapter 11

# Funding Mechanism

The ambitious vision of the **National Garbuz Space Academy** and the global network of **Garbuz Space School Academies** requires a sustainable and innovative funding model. Recognizing the need for consistent financial support to establish and maintain these academies worldwide, the funding mechanism draws on resources generated by **Georgiy Sergeyevich Garbuz's** scientific breakthroughs and space-related projects, as well as revenue streams tied to the **International Intergalactic Space Federation (IISF).**

## Leveraging Taxes from Georgiy Sergeyevich Garbuz's Scientific Works and Space-Related Projects

1. **A Legacy of Innovation**:
   • Georgiy Sergeyevich Garbuz's groundbreaking contributions to science, medicine, and space technology have generated substantial revenue streams through patents, licensing, and international collaborations.
   • These funds serve as a cornerstone of the academy's financial model, directly supporting its mission to provide education to millions of students.

2. **Targeted Tax Allocation**:
   A portion of the revenue from Garbuz's scientific works is allocated specifically for educational initiatives. This includes taxes and royalties from:

- **Medical Breakthroughs**: Revenues from life-saving treatments and regenerative medicine therapies pioneered by Georgiy.
- **Resource Utilization Innovations**: Earnings from space mining technologies and sustainability-focused engineering solutions.

3. **Philanthropic Commitment**:

- Georgiy's personal dedication to education drives the reinvestment of his earnings into the academy network. This philanthropic approach ensures that the initiative remains well-funded and aligned with its mission of accessibility and equity.

4. **Scaling with Success**:

- As Georgiy's scientific contributions continue to expand, so does the financial base for the academies. Revenue growth from new discoveries and technologies ensures long-term sustainability and scalability.

# Redirecting Revenue from IISF Space Programs to Sustain and Grow the Academy Network

1. **IISF Revenue Streams**:

The **International Intergalactic Space Federation** generates significant revenue from its interstellar programs and projects, including:

- **Space Exploration Missions**: Profits from asteroid mining, lunar resource extraction, and planetary colonization initiatives.
- **Aerospace Manufacturing**: Earnings from the development and sale of advanced spacecraft, propulsion systems, and exploration equipment.
- **Space Tourism**: Revenue from commercial space travel and orbital tourism ventures.
- **Space-Based Services**: Income from satellite communications, space manufacturing, and renewable energy projects.

2. **Revenue Allocation for Education**:

- A dedicated percentage of IISF's annual revenue is redirected to support the academy network. This allocation is structured to ensure that both the operational needs of the academies and their future expansion are fully funded.

3. **Global Equity Through IISF Contributions**:
   • IISF-member nations contribute to the funding pool, ensuring that academies in developing countries receive the resources they need to thrive.
   • This equitable distribution reflects the IISF's commitment to fostering global unity and inclusivity.

4. **Reinvestment in Growth**:
   • Revenue from IISF space programs is reinvested into the academy network to expand infrastructure, upgrade technology, and develop new programs that align with emerging trends in space exploration.

5. **Economic Synergy**:
   • The academy network, by preparing a skilled workforce for IISF projects, contributes to the success of space programs. This creates a **cycle of reinvestment**, where educational initiatives and space exploration reinforce each other's growth.

## A Vision of Sustainable Funding

The funding mechanism for the Garbuz Space School Academies exemplifies an innovative and sustainable approach to financing global education. By leveraging the revenue from **Georgiy Sergeyevich Garbuz's** scientific achievements and IISF's space programs, the academies ensure financial stability and long-term growth.

This model is not only about funding schools but about creating a self-sustaining ecosystem that ties education, innovation, and exploration together. It ensures that as humanity progresses in its interstellar ambitions, no child is left behind, and every nation has the opportunity to benefit from the boundless possibilities of space.

# Part IV

## Impact on the Space Industrial Revolution

# CHAPTER 12

# PREPARING THE GLOBAL WORKFORCE FOR SPACE

The **Garbuz Space School Academies** are more than educational institutions; they are the foundation of humanity's preparation for a future among the stars. As space exploration transitions from a distant dream to an imminent reality, the academies aim to equip a new generation with the skills, knowledge, and experience needed to drive this interstellar expansion. From establishing extraterrestrial colonies to advancing medical care for life in space, the graduates of these academies will fuel industries critical to humanity's success beyond Earth.

## Space Colonization: Building a New Frontier

Humanity's future on the Moon, Mars, and beyond depends on a skilled workforce capable of transforming barren landscapes into thriving colonies. Graduates of the Garbuz Academies will play pivotal roles in establishing and sustaining extraterrestrial settlements.

1. **Engineering and Infrastructure Development**:
   • Graduates will design and construct habitats that can withstand extreme conditions, such as the intense radiation and thin atmospheres of Mars or the Moon.

• Innovations in resource utilization, such as using Martian regolith for building materials, will be driven by the expertise cultivated in these academies.

2. **Life-Support Systems**:
• Specialists trained in environmental engineering and sustainable technologies will develop and maintain life-support systems, ensuring breathable air, clean water, and stable temperatures in extraterrestrial colonies.
• These systems will also address waste recycling and energy efficiency, enabling long-term sustainability.

3. **Space Farming and Food Production**:
• Graduates with expertise in space biology and agricultural sciences will create self-sustaining food systems, including hydroponics and aquaponics, to support growing populations on other planets.

4. **Mission Operations and Coordination**:
• The academies will train professionals to manage and coordinate complex interplanetary missions, from initial exploration to the ongoing governance of colonies.
• This includes astronauts, mission planners, and interstellar logisticians who can navigate the challenges of operating across vast distances.

## Asteroid Mining and Space Manufacturing

The future of the space economy lies in the vast untapped resources of asteroids and the ability to manufacture goods in space. The Garbuz Academies will prepare a workforce to unlock this potential, creating new industries that will benefit both space-based operations and life on Earth.

1. **Asteroid Mining Specialists**:
• Graduates will develop and operate robotic systems for mining asteroids rich in valuable materials, such as platinum, gold, and water ice.
• They will also refine techniques for extracting and processing resources in microgravity, enabling the production of metals, fuels, and life-support com-

ponents.

2. **Space Manufacturing Experts**:
• The academies will train engineers and technicians to design and operate manufacturing facilities in orbit or on planetary surfaces.
• These facilities will produce high-precision materials, such as ultra-pure semiconductors and advanced composites, which are more efficiently made in zero-gravity environments.

3. **Logistics and Resource Management**:
• Students will learn to manage the complex logistics of transporting resources between asteroids, space stations, and planetary colonies.
• They will also develop sustainable supply chains to ensure the continuous flow of materials needed for construction, exploration, and scientific research.

4. **Economic Impact on Earth**:
• Graduates of the academies will help bridge the gap between space resources and terrestrial markets, driving down costs for rare materials and fostering economic growth worldwide.

## Medical Advancements for Life in Space

Space presents unique challenges for human health, from prolonged exposure to microgravity to increased radiation risks. The Garbuz Academies will prepare the next generation of space medical professionals to address these challenges, ensuring the safety and well-being of astronauts and colonists.

1. **Space Medicine Practitioners**:
• Graduates trained in space medicine will focus on diagnosing and treating health conditions specific to space environments, such as muscle atrophy, bone density loss, and cardiovascular changes.
• They will also develop protocols for emergency medical care during long-duration missions, including surgical procedures in zero-gravity.

2. **Radiation Protection and Mitigation**:

• Researchers will study the effects of cosmic radiation on human biology and develop protective technologies, such as radiation-shielded suits and habitats.

• They will also explore pharmaceutical interventions to reduce long-term health risks associated with radiation exposure.

3. **Psychological Support and Mental Health**:

• Specialists in space psychology will address the mental health challenges of isolation, confinement, and long-term separation from Earth.

• They will design strategies for maintaining morale, reducing stress, and fostering social cohesion in space crews and colonists.

4. **Biotechnology and Space Pharmaceuticals**:

• Researchers will develop biotechnologies to enhance human adaptability in space, including genetic modifications and advanced prosthetics.

• They will also produce pharmaceuticals in microgravity, taking advantage of the unique environment to create purer and more effective drugs.

The graduates of the **Garbuz Space School Academies** will be at the forefront of humanity's transition into a spacefaring civilization. By equipping students with the skills and knowledge to excel in industries like space colonization, asteroid mining, and space medicine, the academies will create a workforce ready to tackle the challenges and opportunities of interstellar exploration.

This new generation of innovators, builders, and caregivers will not only fuel the growth of the space economy but also inspire humanity to dream bigger and achieve more than ever before. Through their contributions, the promise of the stars will become a reality—not just for a few, but for all.

# CHAPTER 13

# DRIVING INNOVATION AND SCIENTIFIC DISCOVERY

The **National Garbuz Space Academy** and its global network of **Garbuz Space School Academies** are not only dedicated to educating the next generation but also to pushing the boundaries of human knowledge and innovation. By serving as hubs for cutting-edge research and development, these academies will contribute ground-breaking technologies that advance both space exploration and life on Earth.

## Spearheading Research for Sustainable Space Exploration

The National Garbuz Space Academy will lead efforts to develop new technologies that make space exploration and colonization more efficient, sustainable, and accessible. Its research programs will focus on solving the unique challenges posed by extraterrestrial environments.

1. **Resource Utilization and Sustainability**:
   • Researchers will pioneer methods for **in-situ resource utilization (ISRU)**, enabling astronauts and colonists to extract and use local resources, such as water ice on the Moon or Martian regolith for construction.

• Innovations in sustainable energy systems, such as advanced solar panels and nuclear reactors, will power habitats and exploration missions in remote and hostile environments.

2. **Advanced Space Propulsion**:
• The academy will collaborate with space agencies and private companies to develop next-generation propulsion systems, such as ion drives and nuclear thermal propulsion, to reduce travel times between planets.
• Efficient propulsion systems will also lower costs, making space exploration more accessible to a wider range of stakeholders.

3. **Life-Support Systems**:
• Research into closed-loop life-support systems will ensure the recycling of air, water, and nutrients, allowing humans to thrive in space for extended periods.
•These systems will be critical for long-term missions, such as colonizing Mars or exploring the outer planets.

4. **Autonomous Technologies**:
• The development of autonomous robotics and AI systems will support exploration missions and colony management, reducing the need for human intervention in hazardous environments.
• These technologies will also enable deep-space missions to operate independently of Earth-based support.

## Innovations for Earth-Based Industries

The technologies and breakthroughs developed in the Garbuz academies will have a profound impact on Earth-based industries, addressing some of humanity's most pressing challenges in energy, healthcare, and infrastructure.

1. **Energy Solutions**:
• Research into advanced solar technologies for space missions will lead to more efficient solar panels for Earth, reducing reliance on fossil fuels and accelerating the transition to renewable energy.

• Innovations in energy storage, such as compact batteries and thermal systems, will have applications in electric vehicles, grid stability, and portable energy solutions.

2. **Healthcare Advancements**:
• Biomedical research conducted to address the health challenges of space exploration, such as muscle atrophy and radiation exposure, will lead to new treatments and therapies for conditions affecting people on Earth.
• Microgravity research will drive advancements in drug development, enabling the production of purer and more effective pharmaceuticals.

3. **Infrastructure Development**:
• Techniques developed for building habitats on the Moon and Mars, such as 3D printing with local materials, will revolutionize construction methods on Earth, reducing costs and environmental impact.
• Modular and prefabricated construction technologies will improve disaster recovery efforts and affordable housing initiatives worldwide.

4. **Agricultural Innovation**:
• Research into sustainable farming techniques for space colonies, such as hydroponics and vertical farming, will improve food security on Earth, especially in regions facing water scarcity or limited arable land.
• These innovations will also reduce the environmental footprint of agriculture by minimizing resource usage and waste.

# A Catalyst for Global Progress

The innovations developed within the Garbuz academies will create ripple effects across multiple sectors, driving economic growth and improving quality of life worldwide.

1. **Economic Impact**:
• Breakthroughs in technology will open new markets, create jobs, and stimulate industries, both in space exploration and on Earth.
• The commercialization of research outcomes will generate revenue streams to

sustain and expand the academy network.

2. **Environmental Benefits**:
  • Sustainable technologies developed for space exploration will reduce humanity's environmental footprint, supporting efforts to combat climate change and resource depletion.
  • Energy-efficient systems and waste-reduction techniques will contribute to a cleaner, greener future.

3. **Global Collaboration**:
  • The research conducted at the academies will foster international partnerships, bringing together scientists, engineers, and policymakers from around the world to solve shared challenges.
  • This spirit of collaboration will enhance humanity's ability to tackle global issues, from pandemics to natural disasters.

By driving innovation and scientific discovery, the **Garbuz Space School Academies** will not only prepare humanity for a future in space but also address critical challenges here on Earth. The technologies developed within these academies will fuel sustainable space exploration, transform industries, and improve lives worldwide.

This dual impact—advancing humanity's reach into the cosmos while enhancing life on Earth—makes the Garbuz academies a cornerstone of global progress, ensuring that their contributions benefit all of humanity, both now and in the generations to come.

# CHAPTER 14

# FOSTERING GLOBAL PEACE AND UNITY

The **Garbuz Space School Academies** stand as a powerful symbol of humanity's collective ambition to reach for the stars. By uniting nations in the shared goal of space exploration, these academies embody the vision of the **International Intergalactic Space Federation (IISF)**: fostering global cooperation, transcending divisions, and building a future grounded in peace and unity.

## A Platform for International Collaboration

Space exploration has long been a unifying force, inspiring nations to work together in pursuit of common goals. The Garbuz Academies take this spirit of collaboration to a new level, creating a global network that brings together students, educators, and scientists from every IISF-member country.

1. **Shared Educational Goals**:
    • By offering a standardized curriculum focused on space sciences, engineering, and health, the academies provide a shared foundation for students from diverse cultural and national backgrounds.
    • This unified approach fosters mutual understanding and respect, as students collaborate on projects and learn from each other's perspectives.

2. **Cultural Exchange and Understanding**:
   • The academies prioritize cultural exchange, encouraging students to celebrate their unique heritage while working toward common objectives.
   • International events, such as space science fairs, student exchange programs, and joint research initiatives, deepen connections between nations and build bridges of understanding.

3. **Diplomacy Through Education**:
   • By involving nations with historically complex relationships, the academies serve as a platform for diplomatic engagement.
   • Collaborative learning and shared achievements help to reduce tensions, fostering trust and cooperation among participating countries.

## Uniting Nations Through the Vision of the IISF

The IISF's vision of global unity through space exploration is deeply embedded in the mission of the Garbuz Academies. By aligning their efforts with IISF goals, the academies contribute to a peaceful and cooperative future.

1. **Global Governance of Space**:
   • The academies emphasize the principles of shared stewardship and responsibility for space, promoting ethical exploration and resource utilization.
   • Students are taught the importance of international treaties, such as the Outer Space Treaty, and the need for cooperation in managing space resources for the benefit of all humanity.

2. **Equitable Access to Opportunities**:
   • By establishing academies in every IISF-member country, the initiative ensures that no nation is excluded from the opportunities of space exploration.
   • This equitable approach strengthens global solidarity, as nations work together to address shared challenges and achieve mutual success.

3. **A Common Future Among the Stars**:
   • The academies prepare students not only for careers in space but also for

roles as global citizens, dedicated to building a future that transcends national boundaries.

• This vision aligns with the IISF's belief that humanity's survival and progress depend on collaboration and unity.

## A Model for Peaceful Collaboration

The Garbuz Academies demonstrate how shared goals can bring nations together, offering a model for peaceful collaboration that extends beyond education.

1. **Joint Space Missions**:
   • The academies will contribute to joint space missions, training students who will work side by side on IISF-led explorations of the Moon, Mars, and beyond.
   • These missions will highlight the power of teamwork, as professionals from diverse backgrounds pool their expertise to overcome challenges and achieve breakthroughs.

2. **Addressing Global Challenges Together**:
   • The skills and knowledge cultivated in the academies will be applied to solving Earth's most pressing issues, from climate change to food security.
   • By working together to address these challenges, nations strengthen their bonds and reaffirm their commitment to a shared future.

3. **Inspiring Unity Through Achievement**:
   • The academies' successes—whether in producing top-tier graduates, advancing scientific research, or contributing to space exploration—will inspire nations to focus on what they can achieve together.
   • These accomplishments will serve as a testament to the potential of collaboration and the strength of shared purpose.

The **Garbuz Space School Academies** are more than institutions of learning; they are catalysts for global peace and unity. By uniting nations in the shared goal of space exploration, the academies embody the IISF's vision of cooperation, fostering a spirit of collaboration that transcends borders and builds trust among nations.

Through education, cultural exchange, and shared achievements, the Garbuz Academies prove that humanity's greatest challenges—and its brightest opportunities—are best approached together. They remind us that as we look to the stars, we must do so not as separate nations, but as a united planet, committed to a future of peace, progress, and unity.

# CHAPTER 15

# ECONOMIC TRANSFORMATION

The **Garbuz Space School Academies** are not only a cornerstone for education and exploration but also a powerful driver of economic growth. By preparing millions of students for careers in the emerging space economy, these academies will spark global job creation and provide participating countries with substantial economic benefits. This transformation will reshape industries, elevate nations, and position humanity for unprecedented prosperity.

## Creation of Millions of Jobs in Space-Related Fields Globally

The space economy is rapidly evolving, driven by advancements in exploration, technology, and resource utilization. The Garbuz Academies are essential in developing a highly skilled workforce capable of meeting the demands of this growing industry.

## Key Sectors for Job Creation:

1. **Space Exploration and Colonization**:
   • Engineers, scientists, and technicians will design and execute missions to the Moon, Mars, and beyond.

Specialists in habitat construction and life-support systems will develop self-sustaining colonies for long-term human habitation.

2. **Aerospace Manufacturing**:
   • Graduates will contribute to the production of spacecraft, satellites, propulsion systems, and other critical technologies.
   • Innovations in lightweight materials and modular designs will further drive growth in this sector.

3. **Space Resource Utilization**:
   • Careers in asteroid mining, lunar resource extraction, and space-based manufacturing will expand rapidly as these technologies mature.

4. **Space Services and Logistics**:
   • Operations management, mission planning, and interstellar supply chain coordination will become essential for maintaining space infrastructure and supporting exploration missions.

## Global Workforce Development:

With millions of students graduating annually, the Garbuz Academies will ensure a steady supply of skilled professionals ready to join the space economy. This workforce will be highly diverse, with talent sourced from every IISF-member country, fostering innovation through collaboration across cultures and disciplines.

## Indirect Job Creation:

The growth of the space economy will also stimulate indirect job creation in supporting industries, such as:

• Renewable energy and sustainable technologies.

• Advanced manufacturing and robotics.

• Education, training, and professional development.

## Economic Benefits for Participating Countries

Nations that invest in the Garbuz Space School Academies and participate in the IISF stand to gain substantial economic advantages, positioning themselves as leaders in the global space economy.

1. **Boosting National Economies**:
   • The development of space-related industries will generate billions in revenue for participating countries.
   • Increased economic activity in sectors like aerospace, mining, and advanced manufacturing will contribute to GDP growth and fiscal stability.

2. **Attracting Investment**:
   • The presence of Garbuz Academies and a skilled workforce will attract investment from multinational corporations, venture capitalists, and international organizations.
   • This influx of funding will spur innovation, infrastructure development, and regional economic revitalization.

3. **Strengthening Technological Leadership**:
   • Participating countries will gain access to cutting-edge research and development, positioning them as leaders in space technologies.
   • Breakthroughs achieved through academy-led projects will enhance global competitiveness and open new markets.

4. **Job Opportunities for Underserved Populations**:
   • By prioritizing education for underserved youth, including those from foster care and marginalized communities, the academies will broaden access to high-paying, high-demand jobs.
   • This approach reduces inequality, empowers local populations, and ensures that the benefits of the space economy are shared equitably.

5. **Revenue from IISF Ventures**:
   • IISF-member countries will receive a share of the revenue generated by inter-

stellar projects, such as asteroid mining, space tourism, and satellite services.

• These funds can be reinvested into national priorities, including education, healthcare, and infrastructure.

## The Ripple Effect: Transforming the Global Economy

The economic transformation driven by the Garbuz Academies extends beyond job creation and national benefits, influencing the global economy in profound ways.

1. **Emergence of a New Industry**:
    • The space economy will become a dominant force in the 21st century, creating new industries and redefining existing ones.
    • The academies will play a central role in this shift, equipping nations to lead and innovate in this rapidly growing sector.

2. **Economic Cooperation Through the IISF**:
    • By fostering collaboration among member countries, the IISF ensures that the benefits of the space economy are distributed equitably.
    • Joint ventures in space exploration and resource utilization will strengthen economic ties and promote global stability.

3. **Sustainable Economic Growth**:
    • Technologies developed for space exploration, such as renewable energy systems and sustainable construction methods, will drive innovation in Earth-based industries.
    • These advancements will contribute to long-term economic growth while addressing global challenges like climate change and resource scarcity.

The **Garbuz Space School Academies** are catalysts for economic transformation, creating millions of jobs, driving innovation, and generating wealth for participating countries. By equipping students with the skills to thrive in the space economy, these academies ensure that nations are prepared to lead in this new frontier.

This vision of economic growth is not limited to individual nations—it is a global effort, fostering collaboration, reducing inequality, and building a future where humanity's shared ambition for the stars translates into prosperity for all.

# Part V

## Social and Humanitarian Impact

# Chapter 16

# Transforming Lives Through Education

The **Garbuz Space School Academies** are founded on a profound belief: education has the power to change lives, uplift communities, and break the cycle of poverty. By prioritizing foster care children and underserved youth, these academies provide access to top-tier education and pathways to rewarding careers. This mission not only transforms individual lives but also fosters a more equitable and inclusive society.

## Offering Opportunities for Foster Care Children and Underserved Youth

Foster care children and underserved youth often face significant barriers to achieving their potential. The Garbuz Academies are dedicated to dismantling these obstacles and providing these children with opportunities they might otherwise never have.

1. **Accessible, High-Quality Education**:
    • The academies offer a world-class curriculum focused on STEM (science, technology, engineering, and mathematics) disciplines, preparing students for careers in high-demand, high-paying fields.
    • Programs are designed to inspire curiosity, creativity, and confidence, fostering a love of learning and discovery among students.

2. **Specialized Support Systems**:
   - Recognizing the unique challenges faced by foster care children, the academies provide targeted support, including mentorship programs, counseling services, and financial assistance.
   - These resources help students build stability and resilience, empowering them to overcome personal and systemic barriers.

3. **Career Pathways in the Space Economy**:
   - By equipping students with skills relevant to space exploration and related industries, the academies open doors to careers as astronauts, engineers, scientists, and medical professionals.
   - Graduates are not only prepared for space-related roles but also possess versatile skills that can be applied across diverse sectors, ensuring long-term career success.

4. **Global Representation and Equity**:
   - By reaching underserved youth in every IISF-member country, the academies create opportunities for children who have historically been excluded from educational and professional advancement.
   - This inclusivity strengthens global talent pools and ensures that the benefits of space exploration are shared equitably.

## Reducing Poverty and Inequality Through Targeted Educational Programs

Education is one of the most effective tools for reducing poverty and inequality. The Garbuz Academies are designed to address these systemic issues through targeted programs that uplift individuals and communities.

1. **Breaking the Cycle of Poverty**:
   - By providing access to high-quality education, the academies empower students to pursue careers that offer financial stability and upward mobility.
   - Graduates are equipped with the skills to secure jobs in the rapidly growing space economy, which offers competitive salaries and long-term career growth.

2. **Closing the Opportunity Gap**:

 • The academies focus on reaching children from marginalized and underserved communities, ensuring that no child is left behind.

 • Scholarships, outreach programs, and partnerships with local organizations help identify and support talented students who might otherwise be overlooked.

3. **Building Stronger Communities**:

 • Graduates often return to their communities as role models and contributors, inspiring others and driving local development.

 • The academies' focus on inclusivity and equity ensures that the benefits of education extend beyond individual students to entire regions.

4. **Economic Empowerment for Families**:

 • When children succeed, their families benefit. The academies provide opportunities for families to break free from cycles of poverty, creating a ripple effect that strengthens communities and economies.

## A Lifeline for Underserved Youth

For foster care children and underserved youth, the Garbuz Academies are more than schools—they are lifelines. They represent a chance to rise above circumstances, achieve dreams, and contribute meaningfully to society.

1. **Transformative Impact on Foster Care Children**:

 • Foster care children often lack stability and resources, but the academies provide an environment where they are valued, supported, and empowered.

 • By giving these children access to education and career opportunities, the academies help them build brighter futures and contribute to breaking systemic barriers.

2. **A Global Movement for Equity**:

 • The Garbuz Academies embody a global commitment to addressing inequality, uniting nations in the shared goal of creating a brighter future for all children.

The **Garbuz Space School Academies** are transforming lives through education by offering foster care children and underserved youth opportunities to access top-tier education and rewarding careers. These academies break the cycle of poverty, reduce inequality, and inspire a new generation to reach for the stars.

Through their commitment to inclusivity, equity, and empowerment, the academies are not just shaping individual futures—they are creating a global movement that uplifts communities, strengthens nations, and ensures a brighter future for all.

# CHAPTER 17

# EQUALITY ACROSS NATIONS

The **Garbuz Space School Academies** are built on the principle that education should be a universal right, not a privilege reserved for the wealthy or the advantaged. By ensuring that every country, regardless of economic standing, has access to these academies and their resources, the initiative promotes global inclusivity and drives progress for humanity as a whole.

## Ensuring Access for Every Country

One of the defining features of the Garbuz Academies is their commitment to reaching every IISF-member country, creating a global network that guarantees no nation is left behind in the pursuit of education and space exploration.

1. **Establishing Academies in All Countries**:
   • The Garbuz Academies will be established in **every city of every IISF-member country**, making advanced STEM education accessible to students in urban centers, remote regions, and underserved communities alike.
   • By tailoring infrastructure and programs to local needs, the academies ensure that even economically disadvantaged countries can fully participate in the initiative.

2. **Equitable Resource Distribution**:
   • Resources, including state-of-the-art technology, curriculum materials, and teacher training, will be distributed equitably across all academies.
   • Financial support from the IISF and wealthier member nations ensures that countries with limited resources can provide the same quality of education as their more affluent counterparts.

3. **Scholarships and Outreach**:
   • Dedicated scholarship programs will prioritize students from economically disadvantaged backgrounds, ensuring that talent and potential are the only factors determining access to education.
   • Outreach efforts will identify and support students in marginalized and remote communities, empowering them to succeed and contribute to the global space economy.

## Promoting Global Inclusivity and Progress

The Garbuz Academies are more than schools—they are instruments of global inclusivity, fostering a shared sense of purpose and ensuring that all nations benefit from advancements in education and technology.

1. **Bridging the Development Gap**:
   • By providing equal access to cutting-edge education, the academies help bridge the gap between developed and developing nations.
   • Graduates from all countries will have the skills and knowledge to participate in and benefit from the growing space economy, creating opportunities for national growth and development.

2. **Fostering Unity Through Education**:
   • Students from diverse backgrounds will collaborate on shared projects, learning to respect and celebrate each other's cultures while working toward common goals.
   • This collaboration builds a sense of global citizenship, preparing students to tackle challenges that transcend borders, such as climate change, resource

scarcity, and space exploration.

### 3. Creating a Global Talent Pool:

• By ensuring every nation contributes to the global workforce, the academies harness the full spectrum of human talent and creativity.

• This diversity of thought and expertise accelerates innovation and ensures that humanity's progress is truly inclusive.

### 4. Economic and Social Progress:

• The academies not only prepare students for high-paying careers in the space economy but also empower them to uplift their families and communities, driving broader economic and social progress.

• Nations with access to Garbuz Academies will see advancements in science, technology, and education, contributing to their overall development.

## A Vision of Equity and Opportunity

The Garbuz Academies are a powerful reminder that global progress depends on inclusivity and equity. By ensuring that every country has access to these transformative resources, the initiative creates a future where no nation is left behind in the pursuit of knowledge, innovation, and exploration.

### 1. Empowering Underserved Nations:

• Countries with limited access to advanced education and technology will benefit immensely from the resources and opportunities provided by the academies.

• This empowerment ensures that the benefits of the space economy are shared equitably, fostering a sense of collective achievement.

### 2. Strengthening Global Collaboration:

• The network of Garbuz Academies serves as a platform for international collaboration, where students, educators, and researchers from all nations work together to solve shared challenges.

• This spirit of cooperation reinforces global unity and creates a foundation for lasting peace and prosperity.

The **Garbuz Space School Academies** embody a commitment to equality across nations, ensuring that every country—regardless of its economic standing—has access to top-tier education and the resources necessary to thrive in the space economy.

By promoting inclusivity and progress, these academies not only transform individual lives but also elevate entire nations, creating a more equitable and united world. Through their global reach and unwavering dedication to opportunity for all, the Garbuz Academies serve as a beacon of hope, proving that humanity's greatest achievements are those we accomplish together.

# Chapter 18

# The Future of Humanity

The **Garbuz Space School Academies** are not just preparing students for the challenges of today; they are equipping a generation to shape the future of humanity. The graduates of these academies will be at the forefront of efforts to build self-sustaining colonies, explore and understand extraterrestrial environments, and push humanity's presence deeper into the cosmos. Their work will lay the foundation for a new era of interplanetary civilization and global progress.

## Building Self-Sustaining Colonies

One of the most significant contributions of Garbuz Academy graduates will be the establishment and maintenance of self-sustaining colonies on the Moon, Mars, and beyond. These colonies will not only serve as stepping stones for deeper exploration but also as examples of sustainable living that can inspire solutions on Earth.

1. **Designing Advanced Habitats**:
    - Engineers and architects trained at the academies will create habitats capable of withstanding harsh extraterrestrial conditions, such as extreme temperatures, radiation, and microgravity.
    - These habitats will incorporate cutting-edge technologies, including 3D

printing with local materials and energy-efficient designs.

2. **Sustainable Life-Support Systems**:

• Specialists in environmental engineering will develop systems to recycle air, water, and waste, creating closed-loop ecosystems that enable long-term habitation.

• Advanced agricultural techniques, such as hydroponics and vertical farming, will provide colonies with fresh food while minimizing resource usage.

3. **Managing Interstellar Communities**:

• Graduates trained in governance and social sciences will establish frameworks for managing interplanetary communities, ensuring harmony, efficiency, and growth.

• These efforts will include developing systems for education, healthcare, and infrastructure in extraterrestrial environments.

## Researching Extraterrestrial Environments

Understanding the environments of other planets, moons, and asteroids is critical for the success of space exploration. Garbuz Academy graduates will lead the way in researching these worlds, uncovering their mysteries, and identifying resources that can support human expansion into space.

1. **Planetary Science and Exploration**:

• Researchers will study the geology, atmosphere, and potential for life on celestial bodies, such as Mars, Europa, and Titan.

• This research will provide valuable insights into the history of the solar system and the conditions necessary for life.

2. **Resource Identification and Utilization**:

• Scientists will locate and analyze resources like water ice, minerals, and metals, determining their potential for supporting human settlements and space industries.

• Techniques for mining and processing these resources will be developed, enabling sustainable operations beyond Earth.

3. **Understanding Space Weather and Radiation**:
 • Graduates will study the impact of solar radiation, cosmic rays, and space weather on both human health and technology, developing protective measures for future missions.

4. **Astrobiology and the Search for Life**:
 • Astrobiologists trained at the academies will search for signs of life on other planets and moons, advancing our understanding of life's potential beyond Earth.

## Expanding Humanity's Reach into the Cosmos

The long-term vision of the Garbuz Academies extends far beyond our solar system. Graduates will be instrumental in laying the groundwork for humanity's exploration of the stars and the establishment of interstellar civilizations.

1. **Next-Generation Spacecraft Development**:
 • Engineers and scientists will design spacecraft capable of traveling vast distances, incorporating advanced propulsion systems like nuclear thermal engines and solar sails.
 • These spacecraft will support both robotic and crewed missions to explore distant planets and star systems.

2. **Interstellar Exploration**:
 • Researchers will lead missions to explore planets orbiting other stars, seeking habitable worlds and understanding the conditions of distant systems.
 • These efforts will expand humanity's knowledge of the universe and potentially identify new homes for future generations.

3. **Establishing Spaceports and Waystations**:
 • Graduates will help build a network of spaceports and waystations throughout the solar system and beyond, enabling efficient travel and logistics for interstellar missions.

- These facilities will serve as hubs for trade, research, and cultural exchange, fostering a sense of unity among interplanetary communities.

4. **Cultural and Scientific Legacy**:
   - As humanity reaches farther into the cosmos, the graduates of the Garbuz Academies will carry with them the values of collaboration, inclusivity, and curiosity, ensuring that exploration is guided by principles of peace and shared progress.
   - Their discoveries will not only enhance our understanding of the universe but also inspire future generations to dream bigger and aim higher.

The future of humanity is inextricably linked to the efforts of those who dare to explore and expand our horizons. The graduates of the **Garbuz Space School Academies** will lead the charge in building self-sustaining colonies, unlocking the secrets of extraterrestrial environments, and pushing humanity's reach into the cosmos.

Their work will not only transform the way we live and explore but also redefine what it means to be human—united by a shared vision, inspired by curiosity, and committed to progress. As humanity steps boldly into the stars, it will do so with the guidance, innovation, and determination of those prepared by the Garbuz Academies.

# CHAPTER 19

# BUILDING A BRIGHTER FUTURE TOGETHER

The **National Garbuz Space Academy** and its global network of **Garbuz Space School Academies** represent more than institutions of learning; they are beacons of hope, progress, and unity. Together, they symbolize humanity's commitment to education as a transformative force that drives innovation, fosters opportunity, and unites the world in a shared vision of a brighter future among the stars.

## A Symbol of Progress and Foundation for the Space Workforce

The **National Garbuz Space Academy,** as the flagship of this ambitious initiative, embodies the pinnacle of educational excellence and global collaboration. Its mission is clear: to prepare a generation that will lead humanity's ventures into space while solving the challenges of life on Earth.

1. **The National Academy as a Global Symbol**:
   • As the largest and most advanced space-focused academy in the world, the National Garbuz Space Academy stands as a symbol of progress, showcasing what humanity can achieve through dedication, vision, and unity.
   • It is the epicenter of research, innovation, and training, setting the standards for excellence in space education and workforce development.

2. **Global Academies as the Foundation**:
- The network of **Garbuz Space School Academies**, established in every IISF-member country, ensures that no talent is left untapped.
- By providing millions of students annually with a rigorous, future-focused education, these academies lay the groundwork for a workforce capable of meeting the demands of the rapidly growing space economy.
- Together, the National Academy and the global academies create a seamless pipeline of knowledge, skill, and innovation, empowering humanity to explore and thrive beyond Earth.

## The Transformative Power of Education

Education is at the heart of this initiative, not just as a means of learning but as a force that transforms individuals, communities, and nations.

1. **Driving Innovation**:
- The Garbuz Academies equip students with the tools and knowledge to solve humanity's most pressing challenges, from developing sustainable technologies to advancing medical care for space exploration.
- The research and discoveries made within these academies ripple outward, sparking progress across industries and inspiring new possibilities.

2. **Creating Opportunity**:
- By prioritizing underserved youth, including foster care children, the academies ensure that education is a gateway to opportunity for those who need it most.
- Graduates are prepared to excel in high-demand, high-paying careers, breaking cycles of poverty and contributing to the global economy.

3. **Fostering Global Unity**:
- The academies bring together students from diverse backgrounds, fostering a sense of shared purpose and collaboration that transcends borders.
- As students learn and grow together, they build the relationships and understanding necessary for global cooperation in addressing the challenges of both

Earth and space.

## Envisioning a Future for All\

The vision of the Garbuz Academies is not confined to the boundaries of our planet; it is a call to imagine a future where every child has the chance to thrive and where humanity ventures into the stars as one.

1. **A Future of Opportunity**:
    • Imagine a world where every child, regardless of their background or circumstances, has access to a world-class education and the tools to achieve their dreams.
    • The Garbuz Academies make this vision a reality, empowering millions of young minds to contribute to humanity's progress.

2. **A Future of Exploration**:
    • Humanity's venture into the cosmos is not the work of a single nation but a collective journey. The academies prepare students to work together to build colonies, explore new worlds, and unlock the secrets of the universe.
    • Their contributions will not only advance our understanding of space but also inspire future generations to dream bigger and reach higher.

3. **A Future of Unity**:
    • As the academies foster collaboration and shared goals, they pave the way for a future where humanity works together, not as divided nations, but as a united species.
    • This unity will be humanity's greatest strength as we face challenges and seize opportunities that lie ahead, both on Earth and beyond.

The **Garbuz Space School Academies** are more than a vision—they are a movement to build a brighter future together. They remind us that education is not just about knowledge but about unlocking potential, creating opportunities, and uniting humanity in pursuit of the extraordinary.

As we look to the stars, let us also look to each other, embracing the power of collaboration and the promise of a future where every child thrives and where humanity's shared dreams become our greatest achievements. Together, through the efforts of the Garbuz Academies and the dedication of the IISF, we can create a world—and a universe—filled with hope, progress, and boundless possibility.

# AFTERWORD

Dear Readers,

At the core of *Garbuz Space School Academies* lies a transformative vision—one where education serves as the bridge to a future of exploration, equality, and boundless opportunity. This book highlights the immense potential we can achieve when we invest in our youth, prioritize inclusivity, and work together to prepare humanity for the challenges and wonders of life among the stars.

Through these pages, we've explored a future where every child, regardless of their background, has access to world-class education and the tools to thrive in the emerging space economy. This book is more than a roadmap; it is a call to action. Every purchase contributes to building the foundation of this vision—establishing the academies that will nurture the next generation of leaders, innovators, and explorers.

By supporting this initiative, you are helping to create a world where talent and potential know no boundaries, where dreams of the cosmos inspire unity on Earth, and where education is the key to unlocking humanity's greatest achievements. Together, we can build a legacy of hope and progress that uplifts all of humanity.

Thank you for joining us on this extraordinary journey. Your belief in this vision makes it possible to transform ambitions into realities—on Earth and beyond. Together, we are charting a course toward a brighter future for every child and a united humanity ready to explore the stars.

# GARBUZ SPACE SCHOOL ACADEMIES

*Proudly sponsored by GarbuzSpace.com*

*Za Detey – For Kids!*

*Za Lyubov – For Love!*

*Live, Make, & Enjoy!*

www.ingramcontent.com/pod-product-compliance
Lightning Source LLC
Chambersburg PA
CBHW071459210326
41597CB00018B/2623